This book Belongs to

_____

_____

_____

To join our mailing list and see other titles available

Website: www.captaintimpublishing.com

Email: info@captaintimpublishing.com

TRACE AND COLOR

TRACE AND COLOR

TRACE AND COLOR

3

TRACE AND COLOR

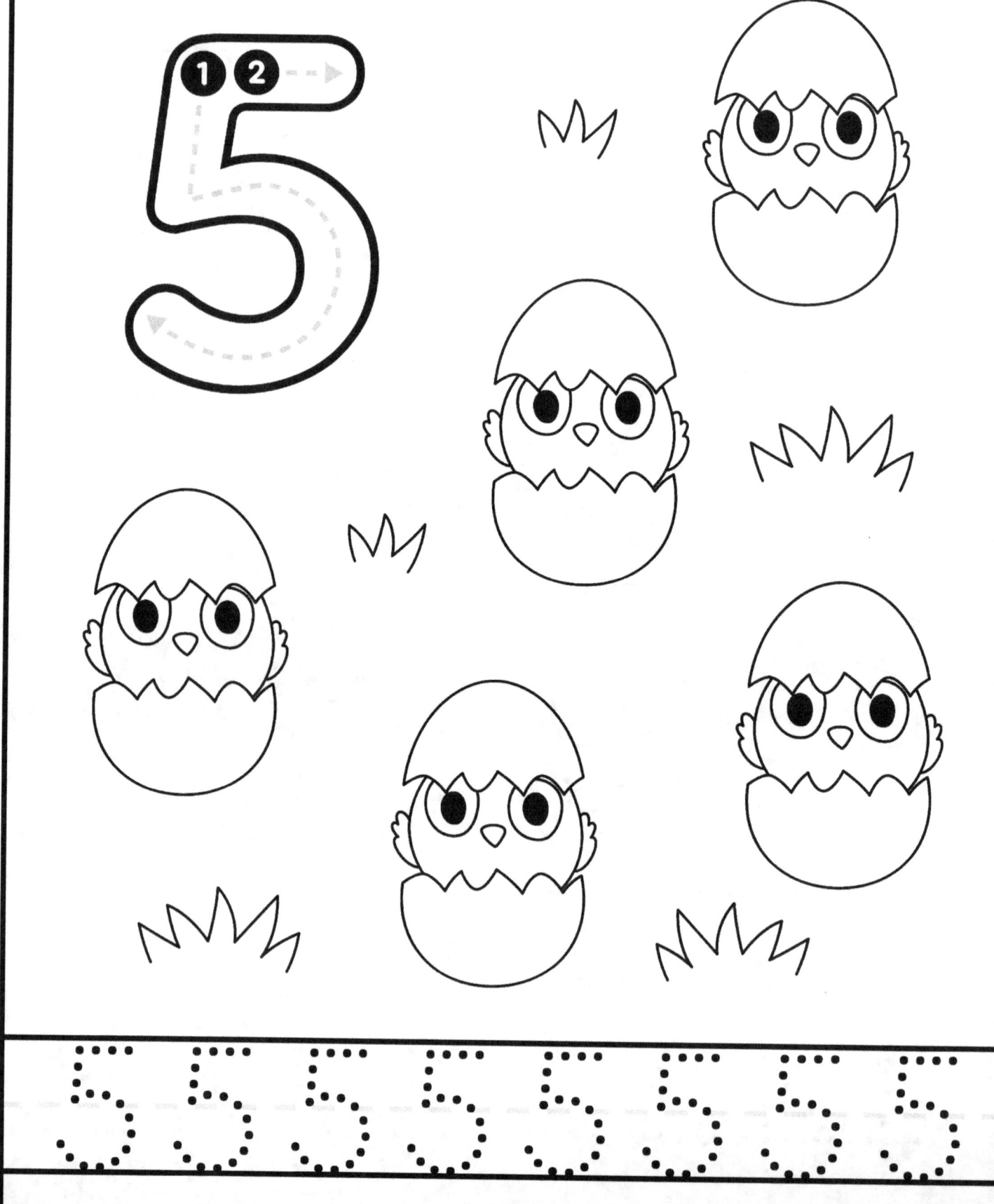

TRACE AND COLOR

TRACE AND COLOR

TRACE AND COLOR

TRACE AND COLOR

SUBTRACTION

Write the correct answer in each box

$3 - 2 = \boxed{1}$

$4 - 1 = \boxed{3}$

$5 - 3 = \boxed{}$

SUBTRACTION

Write the correct answer in each box

 $6 - 3 = \boxed{3}$

 $4 - 3 = \boxed{1}$

 $2 - 1 = \boxed{}$

SUBTRACTION

Write the correct answer in each box

$3 - 1 = \boxed{2}$

$8 - 5 = \boxed{3}$

$9 - 7 = \boxed{}$

SUBTRACTION

Write the correct answer in each box

 $5 - 1 =$ 4

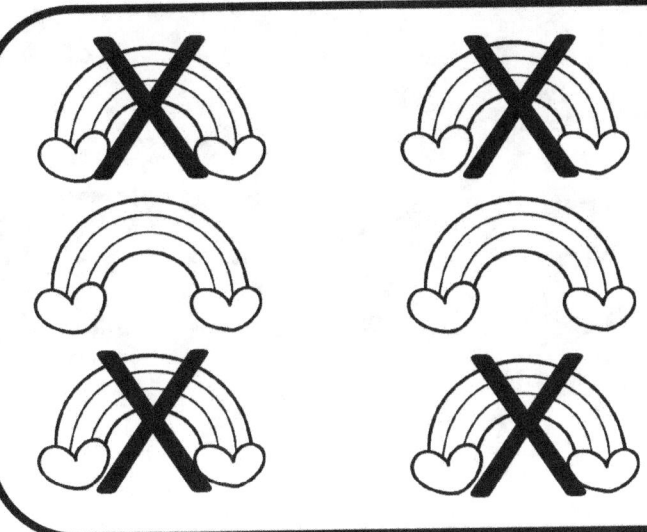 $6 - 4 =$ 2

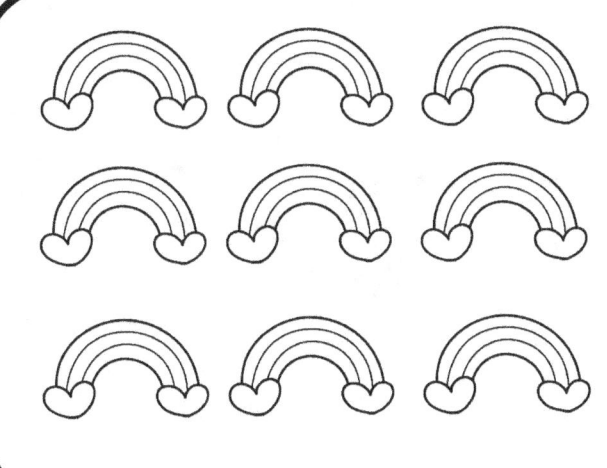 $9 - 8 =$ ☐

SUBTRACTION

Write the correct answer in each box

$1 - 1 = \boxed{0}$

$5 - 2 = \boxed{3}$

$8 - 2 = \boxed{}$

ADDITION

Write the correct answer in each box

4 + 4 = 8

3 + 6 = 9

6 + 2 =

5 + 4 =

ADDITION

Write the correct answer in each box

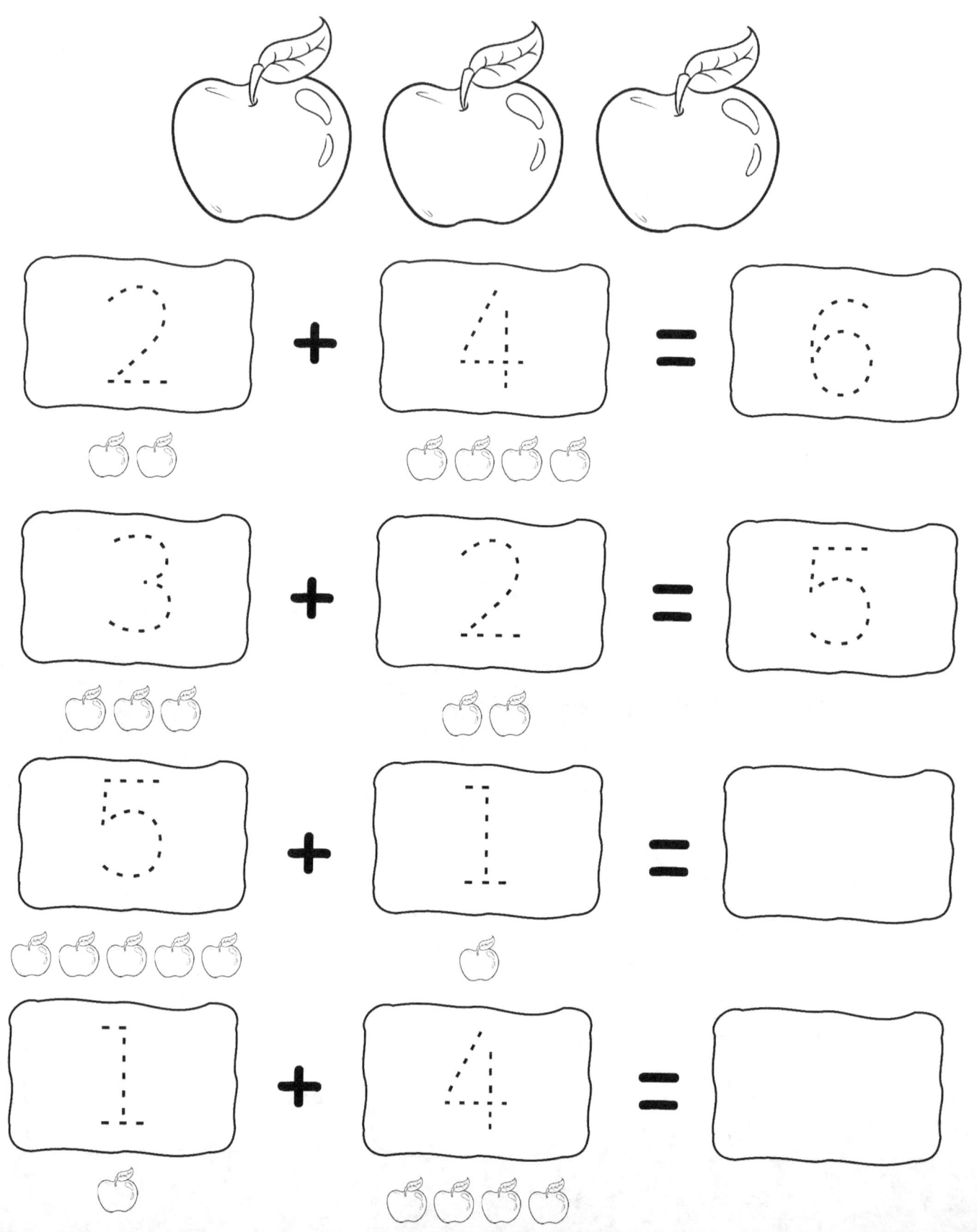

$2 + 4 = 6$

$3 + 2 = 5$

$5 + 1 =$

$1 + 4 =$

ADDITION

Write the correct answer in each box

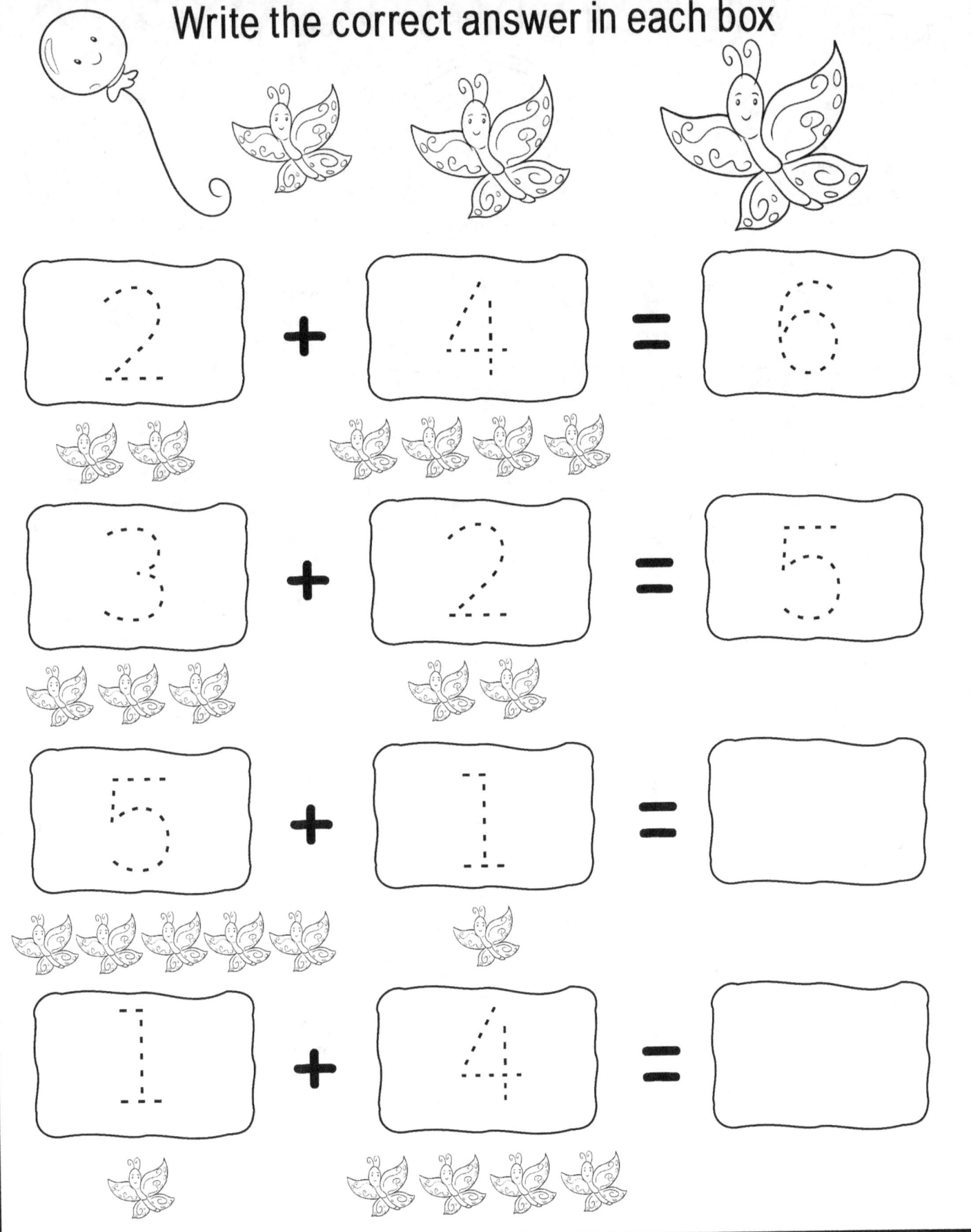

2 + 4 = 6

3 + 2 = 5

5 + 1 =

1 + 4 =

ADDITION

Write the correct answer in each box

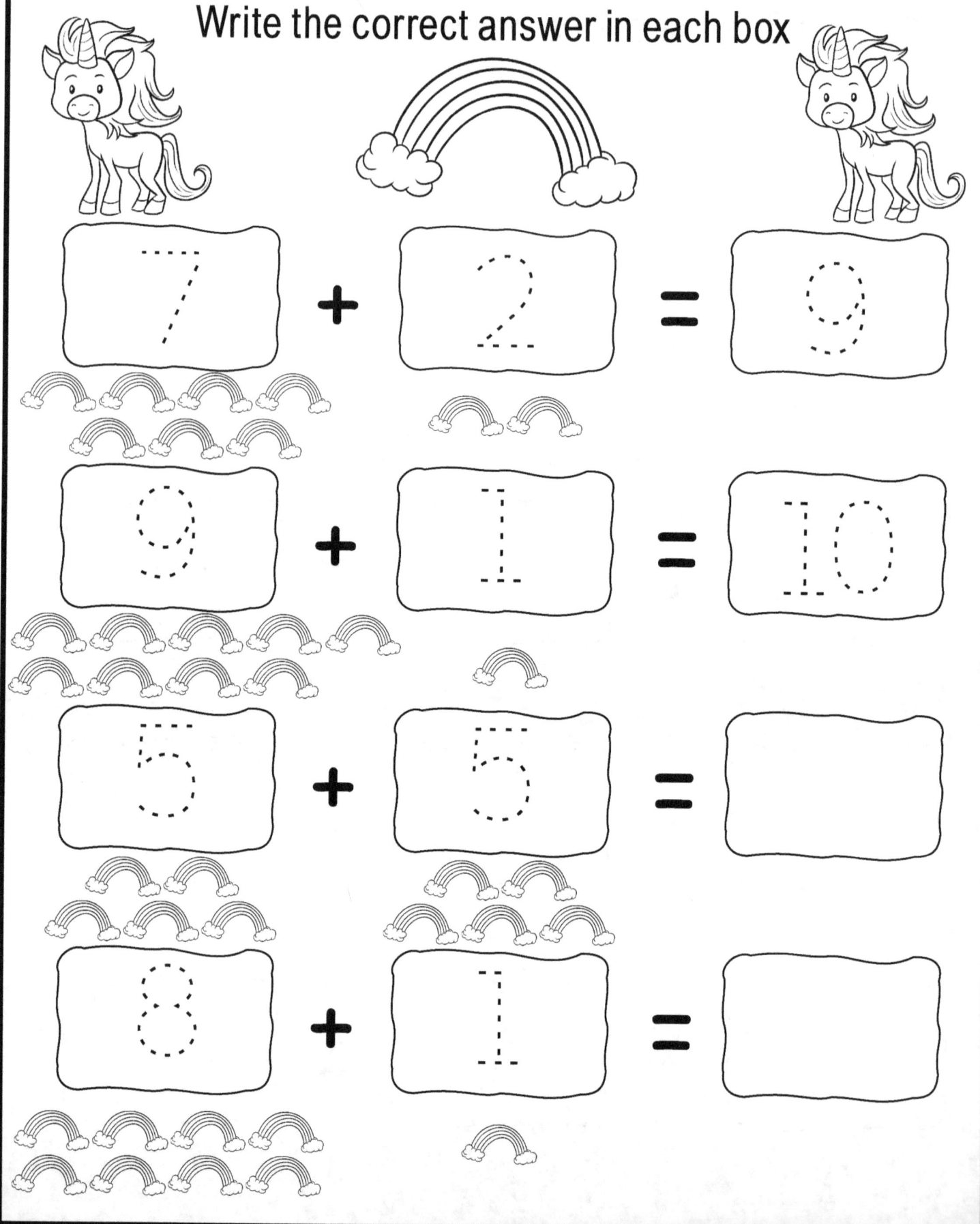

7 + 2 = 9

9 + 1 = 10

5 + 5 =

8 + 1 =

ADDITION

Write the correct answer in each box

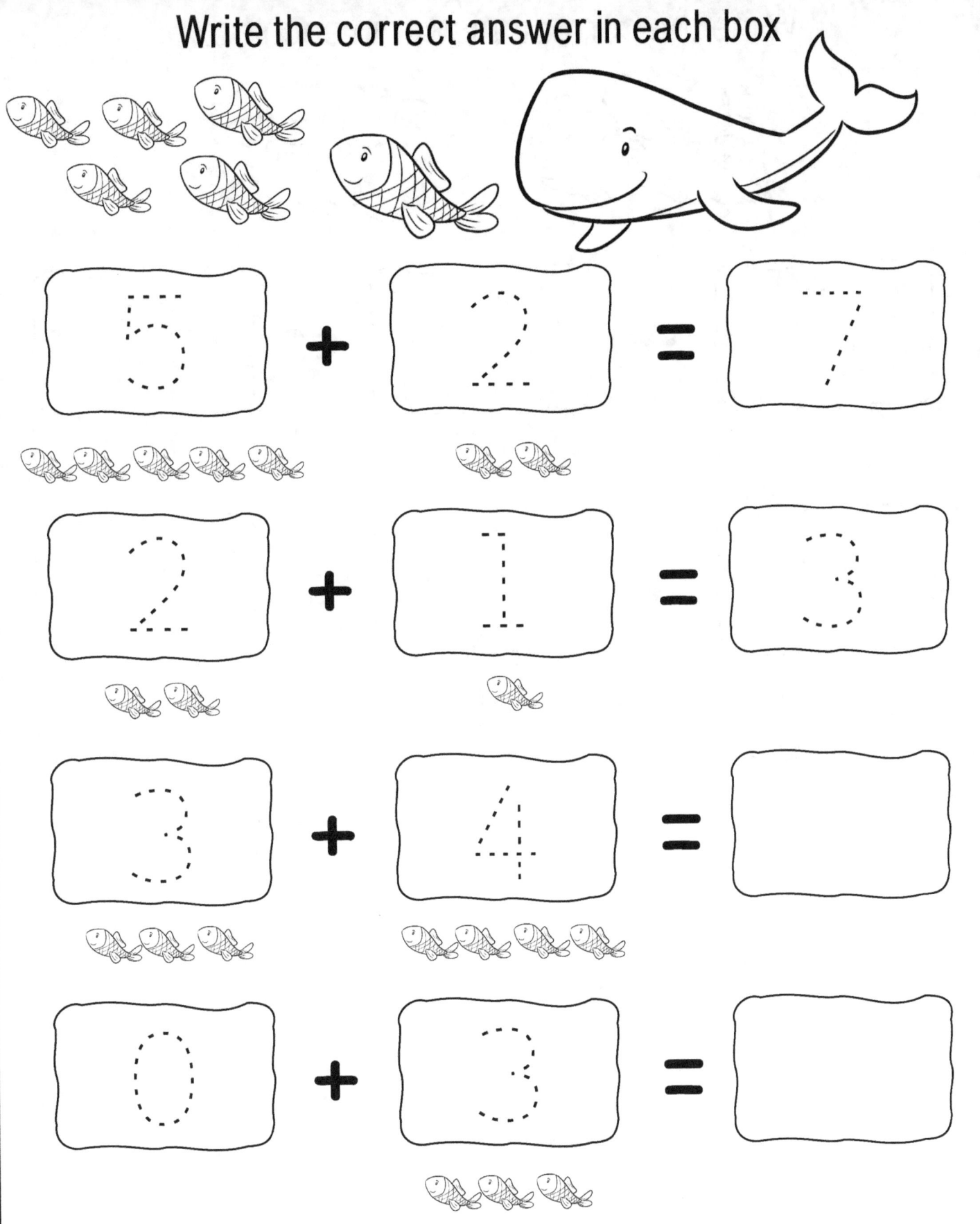

5 + 2 = 7

2 + 1 = 3

3 + 4 =

0 + 3 =

ADDITION

Write the correct answer in each box

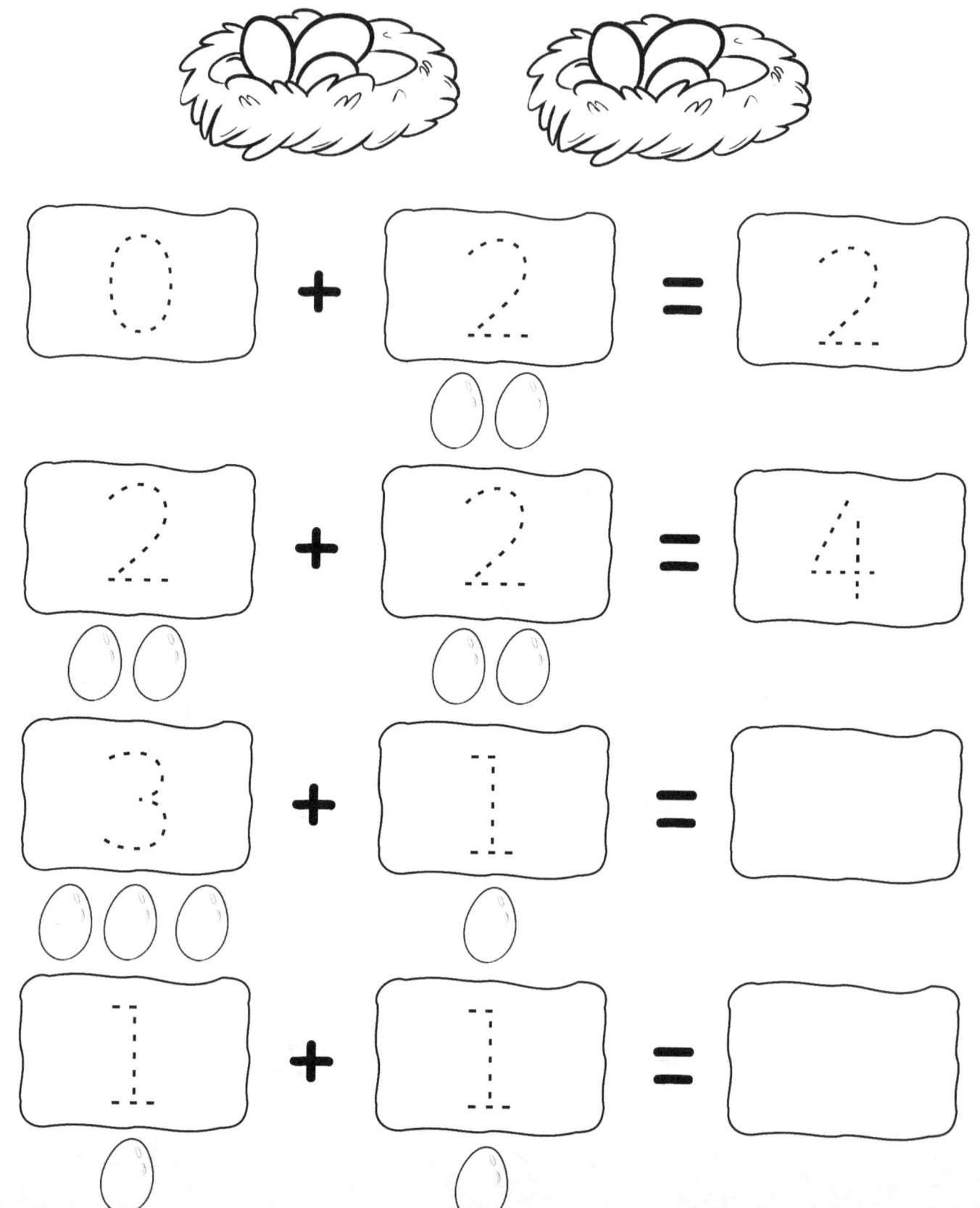

0 + 2 = 2

2 + 2 = 4

3 + 1 =

1 + 1 =

Who is taller?

Color the tallest object in each box

Who is taller?

Color the tallest object in each box

Who is taller?

Color the tallest object in each box

Who is taller?

Color the tallest object in each box

Who is taller?

Color the tallest object in each box

MATCHING

Trace the numbers and match with the pictures

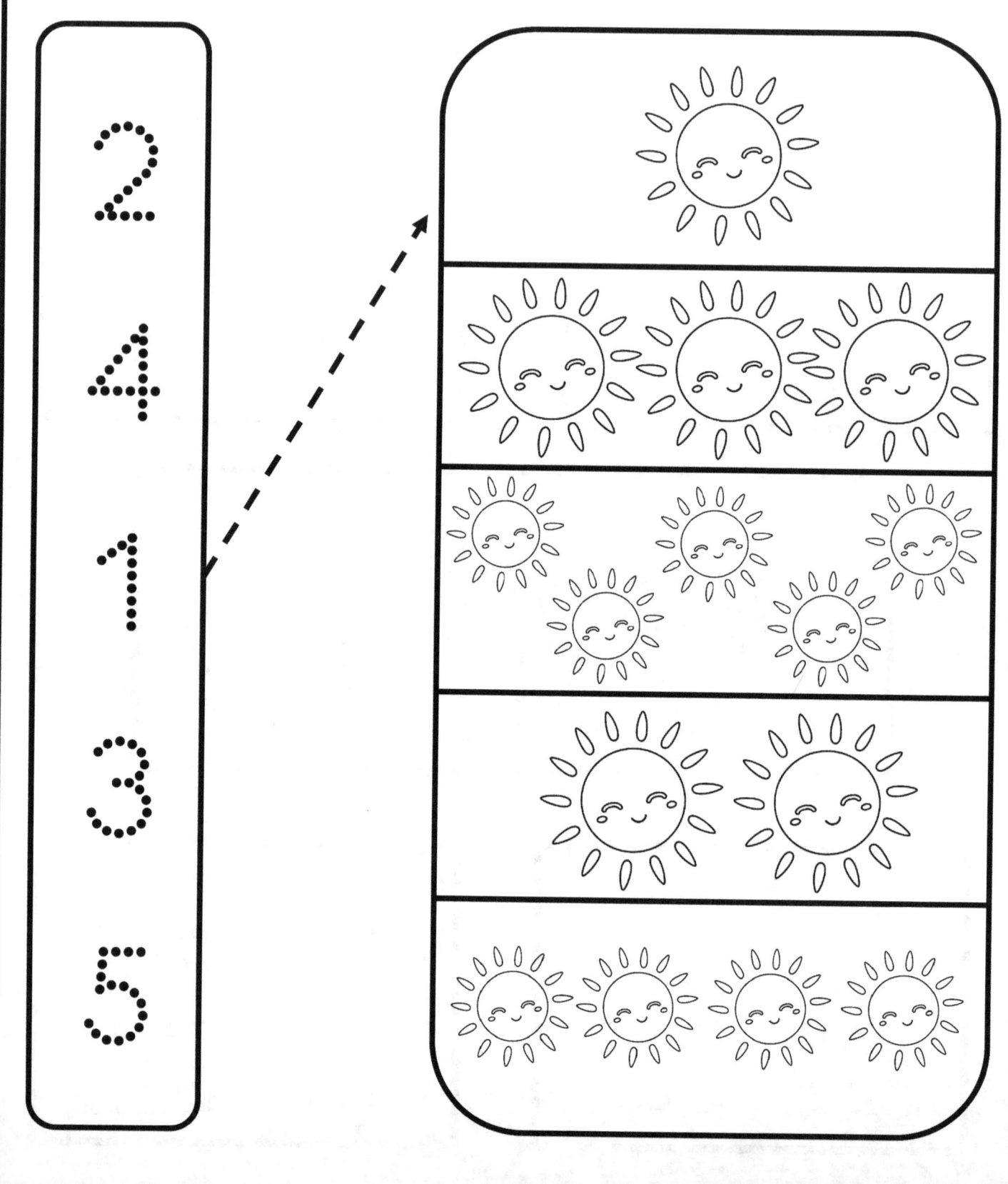

MATCHING

Trace the numbers and match with the pictures

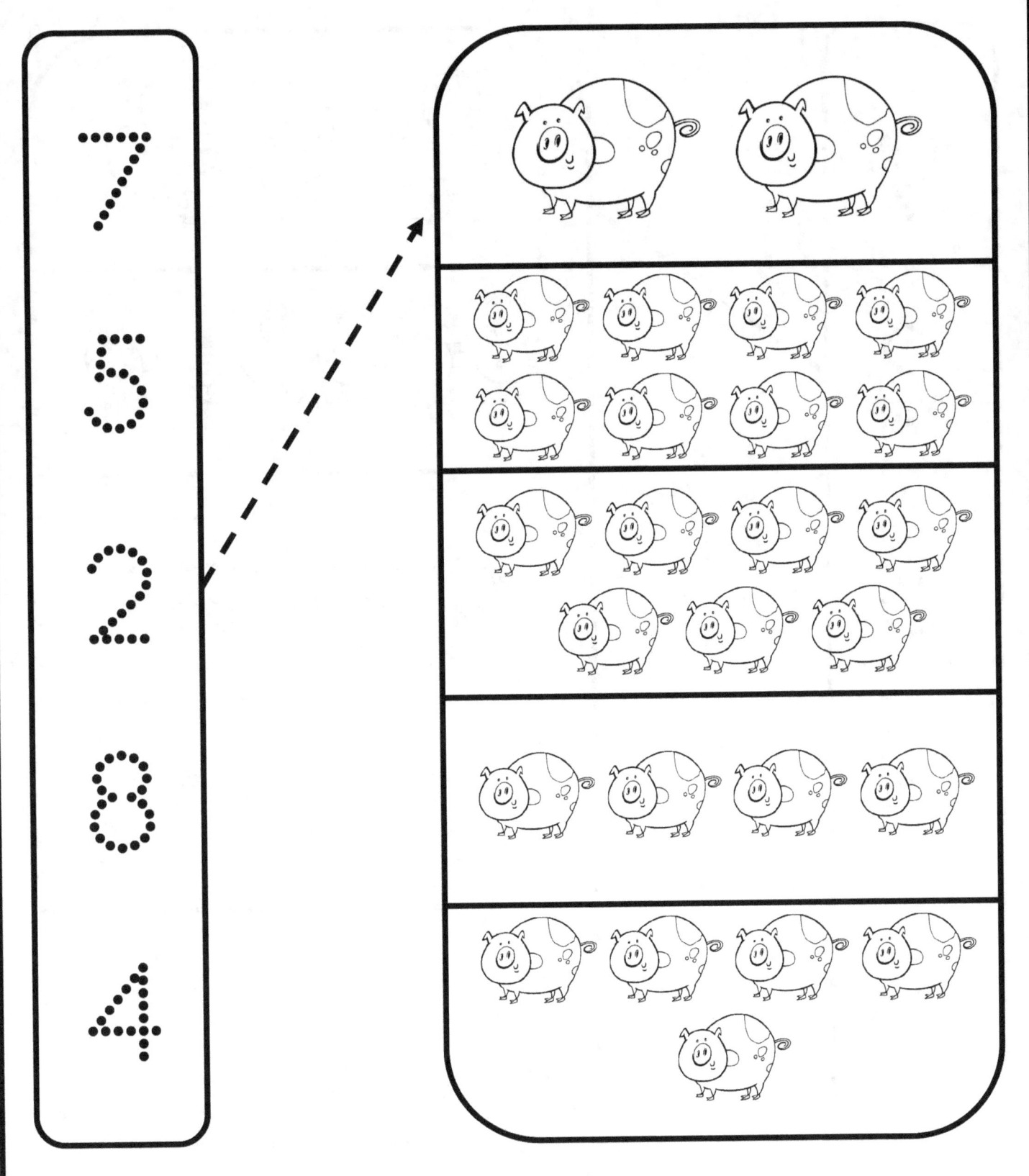

MATCHING

Trace the numbers and match with the pictures

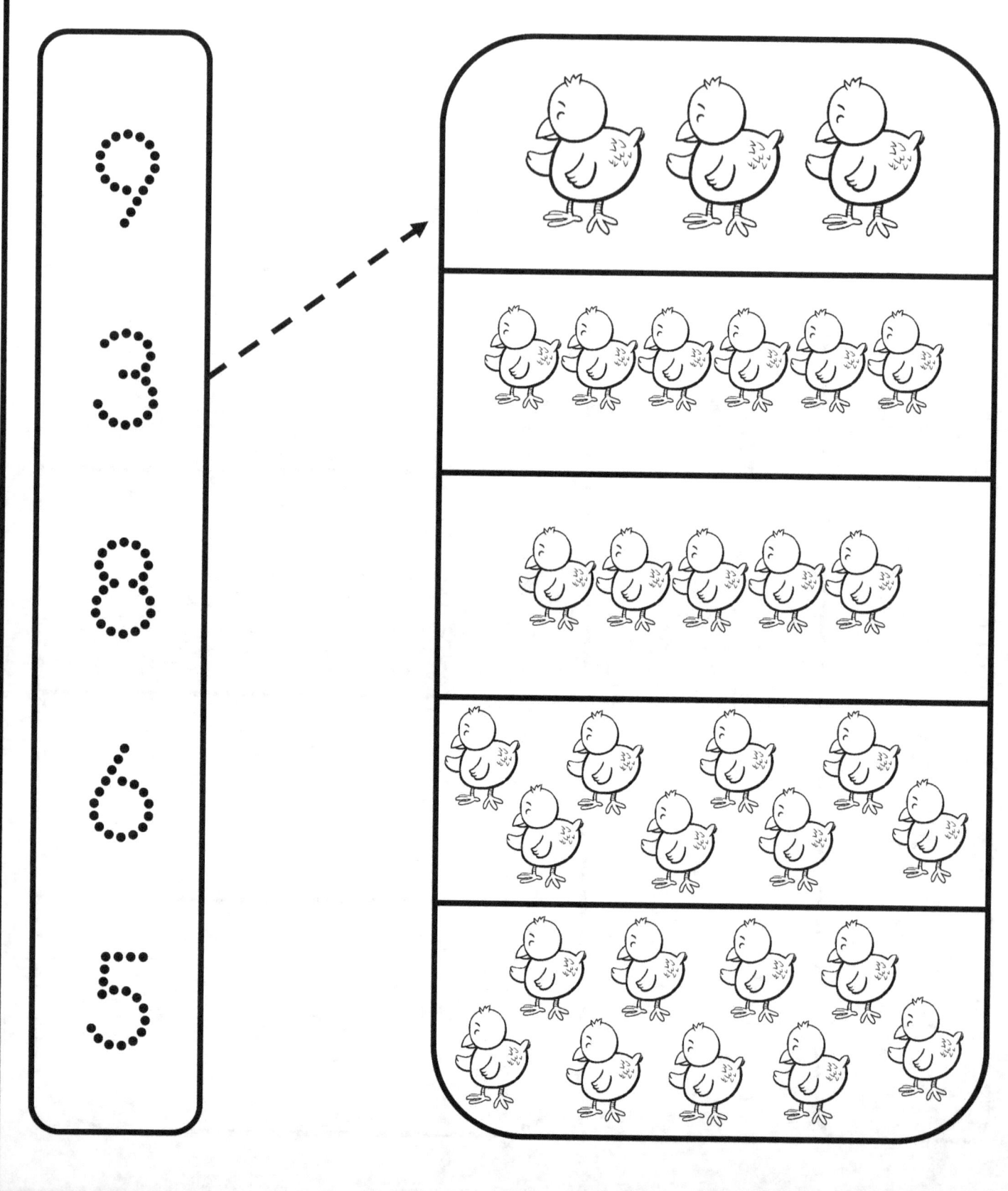

Trace the numbers and match with the pictures

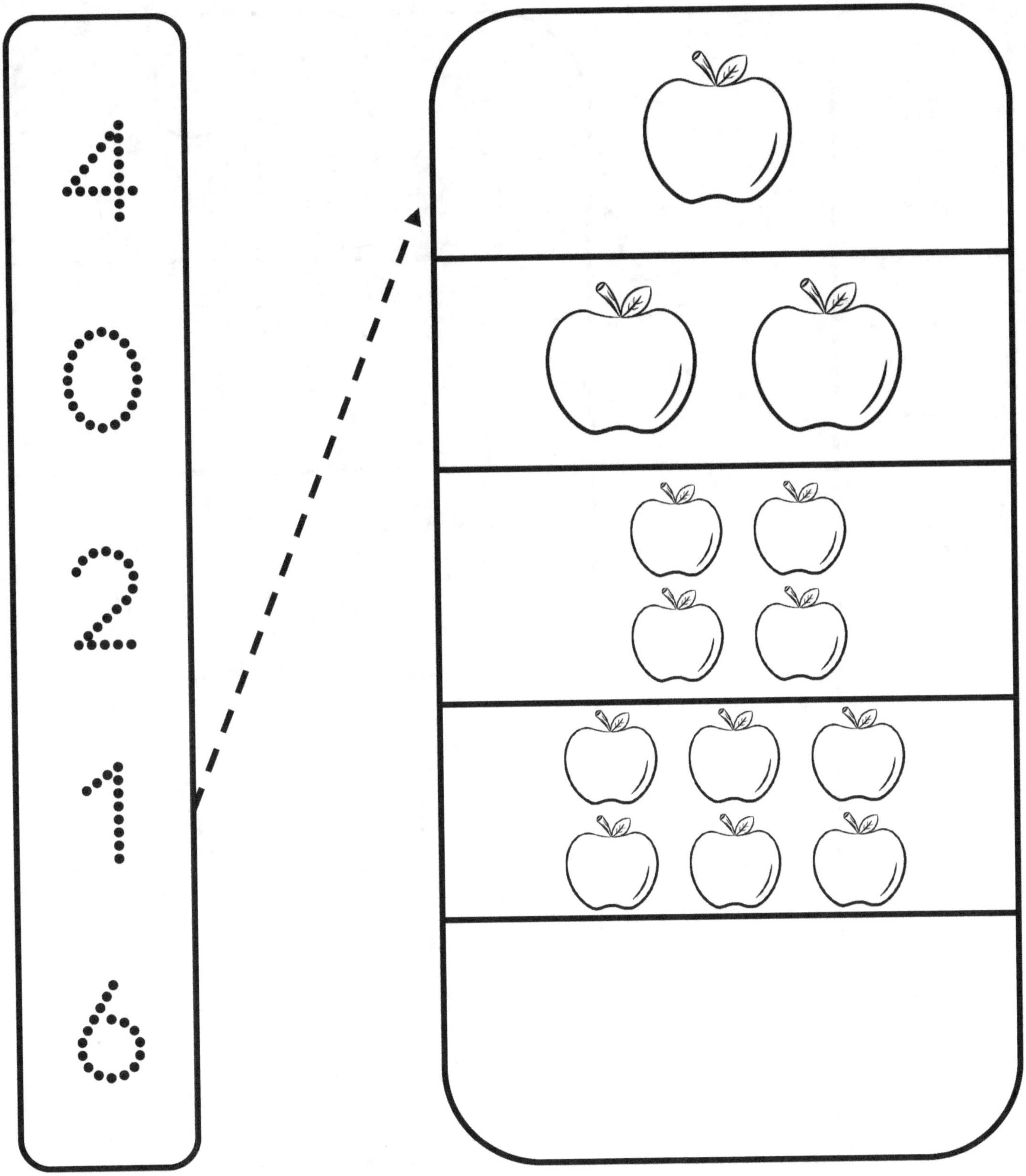

4

0

2

1

6

MATCHING

Trace the numbers and match with the pictures

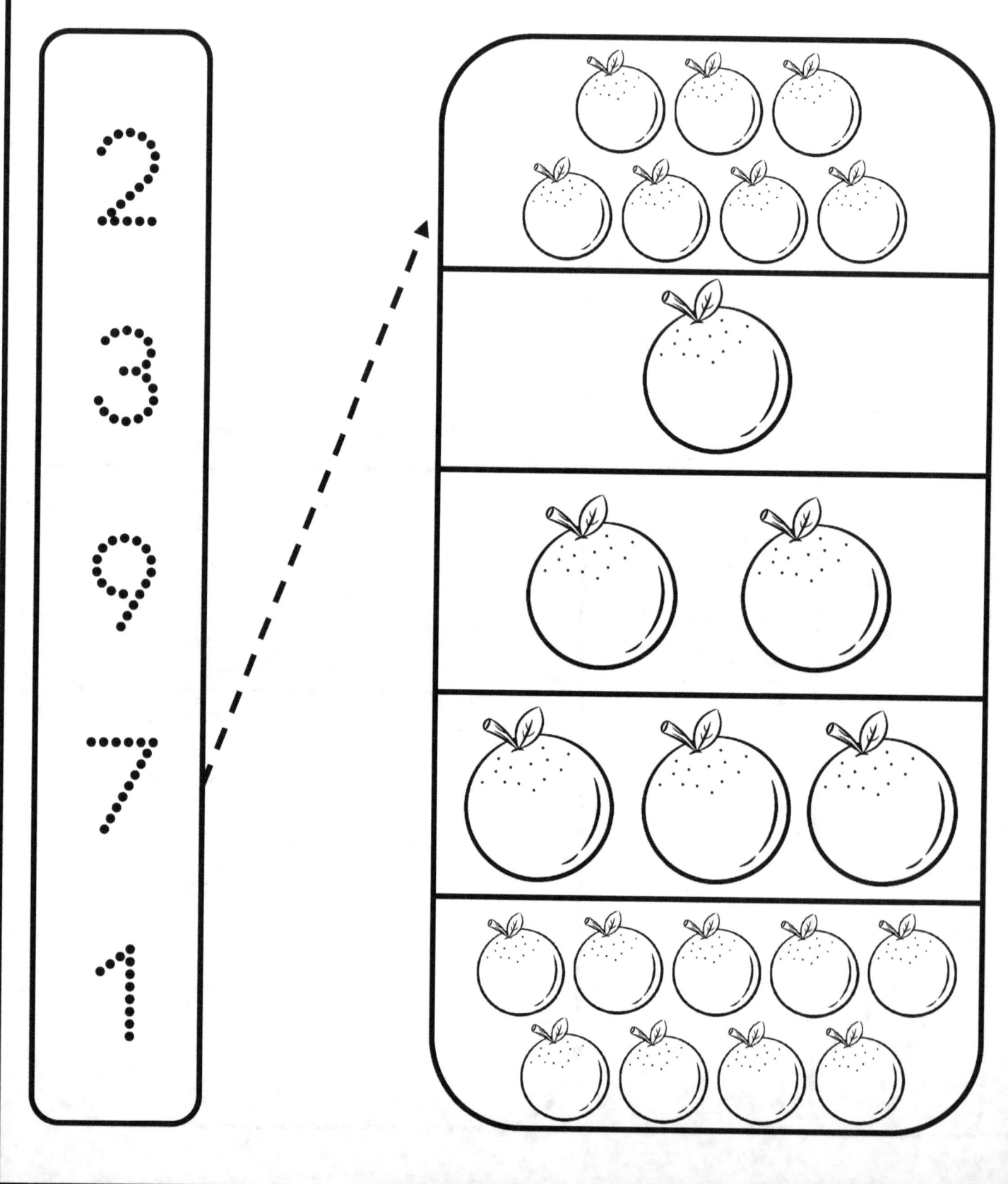

COUNTING

Count the dinosaurs and color the correct number

	3	2	4
	0	5	1
	4	0	5
	2	1	6

COUNTING

Count the apples and
color the correct number

🍎🍎🍎 🍎🍎	2	5	6
🍎🍎🍎🍎 🍎🍎	6	7	5
🍎🍎🍎🍎 🍎🍎🍎🍎	8	2	0
🍎🍎🍎🍎 🍎🍎🍎	2	7	3

COUNTING

Count the ice creams and
color the correct number

	3	7	5
	2	3	1
	4	2	1
	4	3	2

COUNTING

Count the octopuses and
color the correct number

	10	1	2
	7	8	9
	5	2	6
	0	4	7

COUNTING

Count the flowers and
color the correct number

	0	3	2
	4	2	3
	3	1	7
	3	4	5

CONNECTING DOTS

Connect the numbers in ascending order
then color the picture

CONNECTING DOTS

Connect the numbers in ascending order
then color the picture

CONNECTING DOTS

Connect the numbers in ascending order then color the picture

1   6

2•

3

4

•5

CONNECTING DOTS

Connect the numbers in ascending order
then color the picture

CONNECTING DOTS

Connect the numbers in ascending order
then color the picture

CONNECTING DOTS

Connect the numbers in ascending order then color the picture

CONNECTING DOTS

Connect the numbers in ascending order
then color the picture

CONNECTING DOTS

Connect the numbers in ascending order
then color the picture

CONNECTING DOTS

Connect the numbers in ascending order
then color the picture

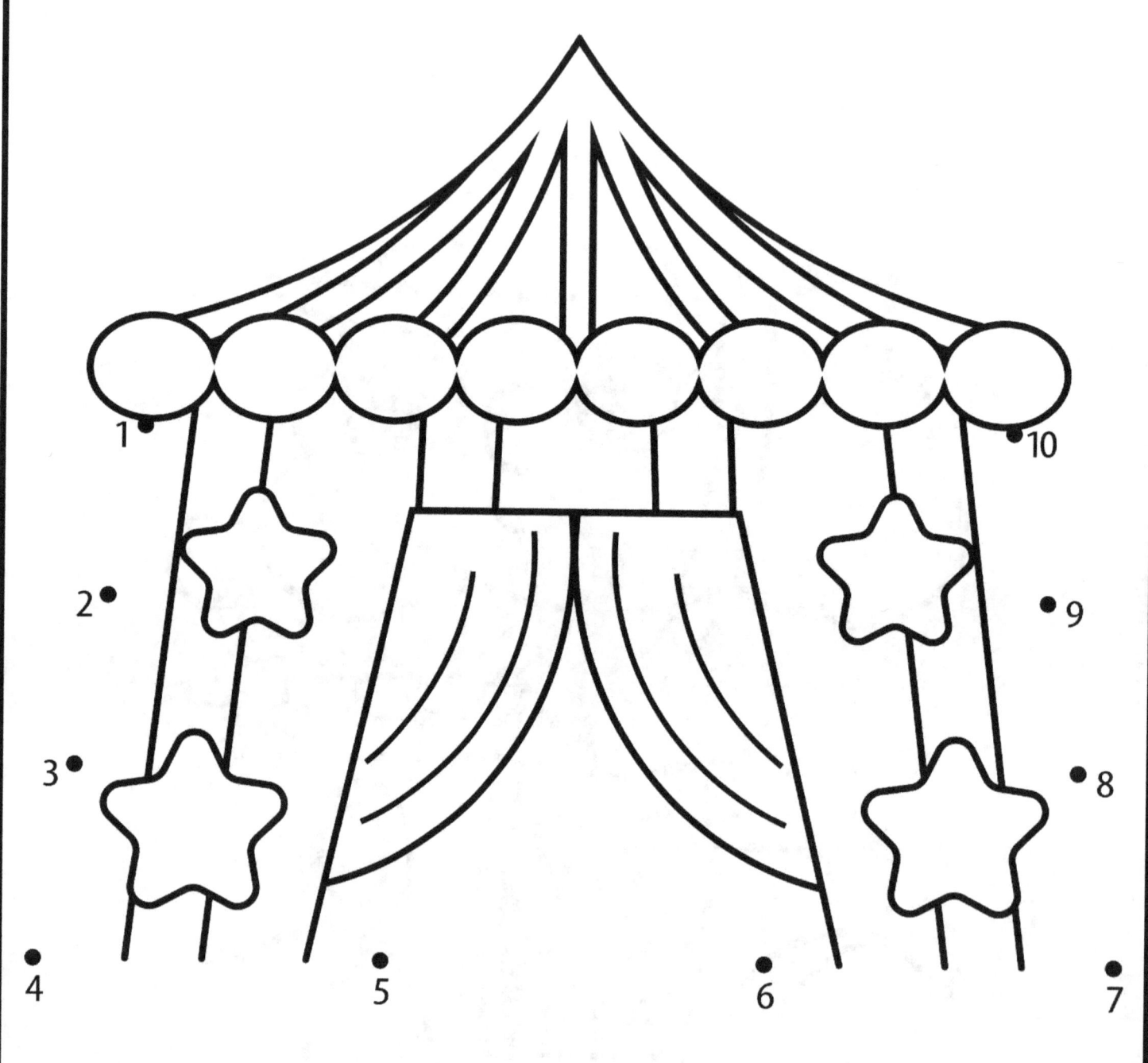

CONNECTING DOTS

Connect the numbers in ascending order then color the picture

COLOR AND TRACE

 Find and color the number 0
Trace the number 0

1	4	3	5
9	0	7	6
4	2	0	1
10	9	8	0

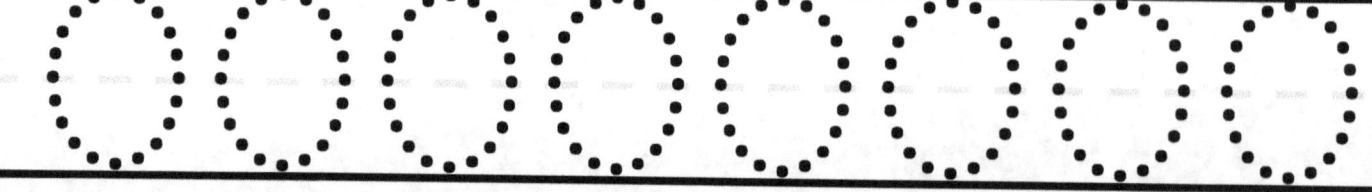

COLOR AND TRACE

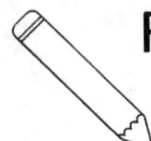

Find and color the number 1

Trace the number 1

0	5	3	2
8	7	1	6
1	3	2	1
10	8	1	0

1 1 1 1 1 1 1 1

COLOR AND TRACE

 Find and color the number 2
Trace the number 2

2	0	4	2
8	9	1	6
2	4	5	2
7	9	1	8

2 2 2 2 2 2 2

COLOR AND TRACE

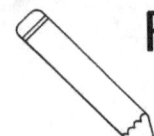

Find and color the number 3

Trace the number 3

3	0	4	2
8	9	2	8
0	6	3	9
5	3	1	10

3 3 3 3 3 3 3 3

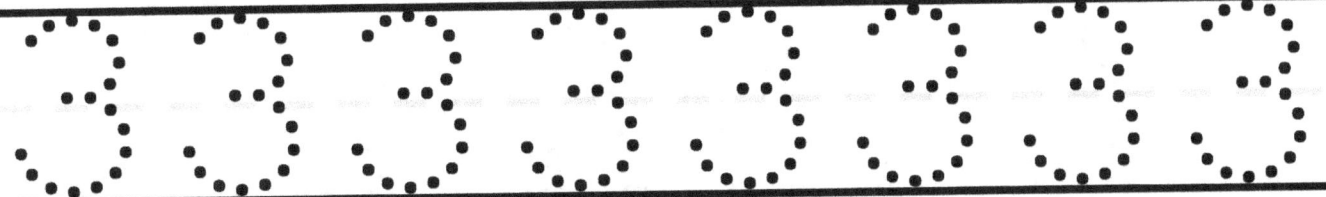

COLOR AND TRACE

 Find and color the number 4
Trace the number 4

4	0	9	2
2	8	2	5
9	6	4	9
4	3	1	10

4 4 4 4 4 4 4

COLOR AND TRACE

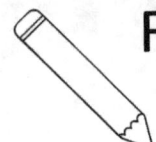 Find and color the number 5
Trace the number 5

7	8	9	5
10	5	2	7
9	6	5	9
5	4	2	10

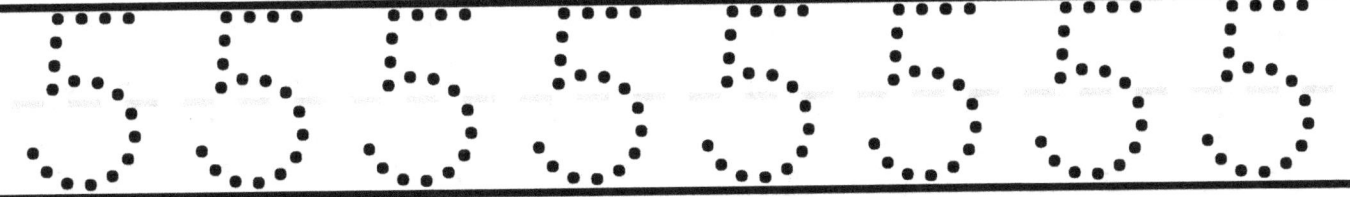

COLOR AND TRACE

 Find and color the number 6
Trace the number 6

2	6	8	2
6	4	1	10
9	8	6	9
6	4	2	10

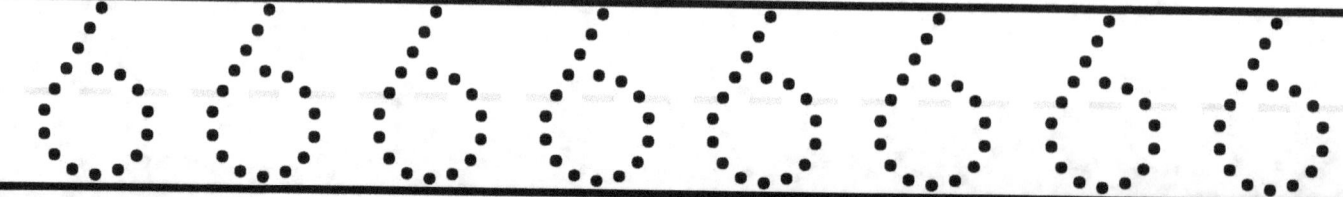

COLOR AND TRACE

 **Find and color the number 7**

Trace the number 7

7	2	8	2
5	4	3	7
0	8	7	9
9	3	2	8

7 7 7 7 7 7 7

COLOR AND TRACE

 Find and color the number 8
Trace the number 8

9	3	5	2
8	6	3	8
8	8	7	9
0	3	1	4

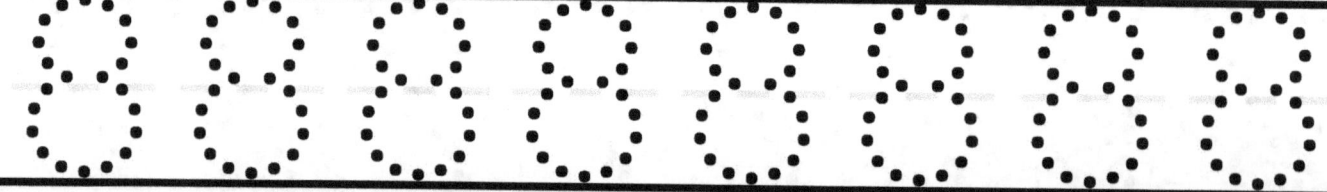

COLOR AND TRACE

 Find and color the number 9
Trace the number 9

9	1	4	9
4	9	3	8
6	5	7	9
9	2	10	4

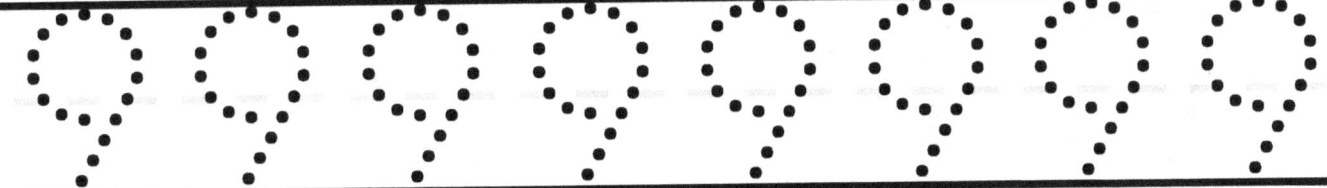

COLOR AND TRACE

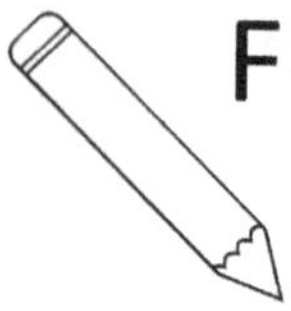

Find and color the number 10
Trace the number 10

5	10	2	1
10	9	3	7
8	5	10	6
3	2	10	4

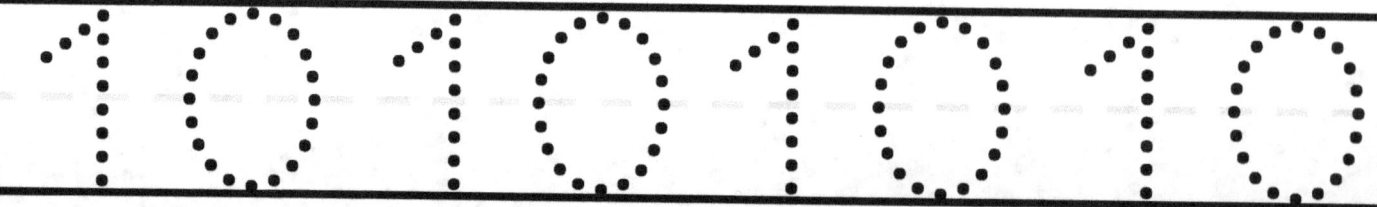

Who is shorter?

Color the shortest object in each box

Who is shorter?

Color the shortest object in each box

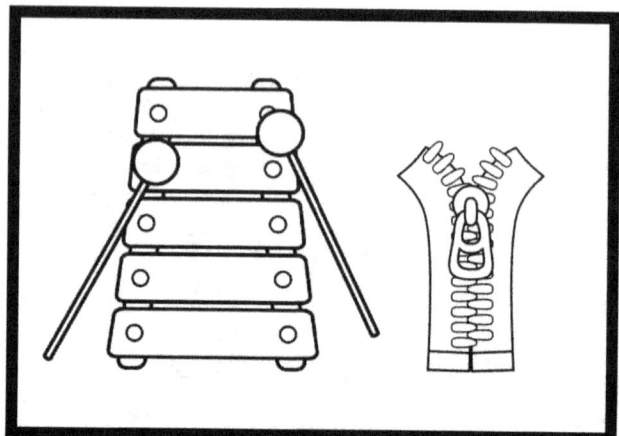

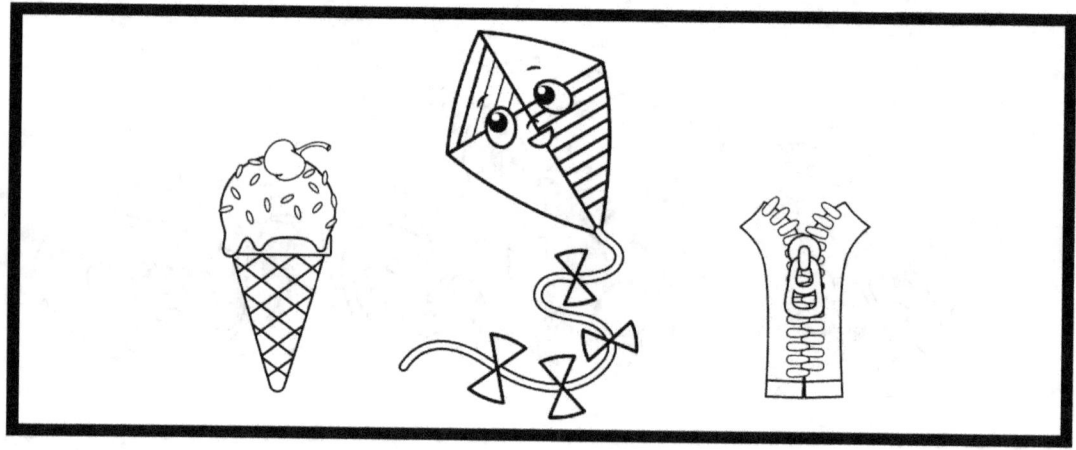

Who is shorter?

Color the shortest object in each box

Who is shorter?

Color the shortest object in each box

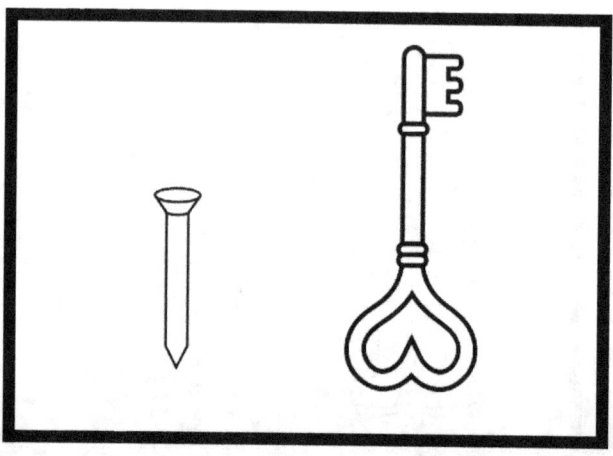

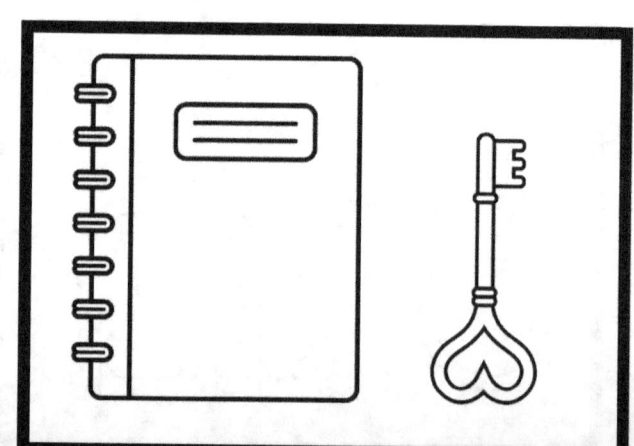

COLOR BY NUMBER
Find and color number 1

COLOR BY NUMBER
Find and color number 2

COLOR BY NUMBER
Find and color number 3

COLOR BY NUMBER

Find and color number 4

4

COLOR BY NUMBER
Find and color number 5

5

6

1

3

5 5 5

8

9

5

5

6

8 5

3

2

5

5

2

1

8

COLOR BY NUMBER
Find and color number 6

6

1

2

9

6

6

3

7

6

6

2

6

8

4

6

4

5

5

COLOR BY NUMBER
Find and color number 7

7

3

7   7

6

2

2

7

4

7

5

7

7

8

9

2

COLOR BY NUMBER
Find and color number 8

COLOR BY NUMBER
Find and color number 9

COLOR BY NUMBER
Find and color number 10

10

Zero

Three

Four

Six

Seven

Eight

Nine

SUBTRACTION

Count, Subtract and write the correct answer

$9 - 1 =$ 8

$7 - 5 =$ 2

$3 - 2 =$ ☐

$8 - 3 =$ ☐

Count, Subtract and write the correct answer

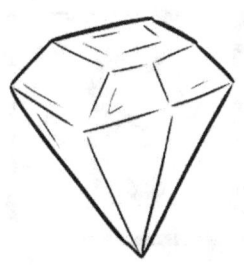 2-1= 1

 5-3= 2

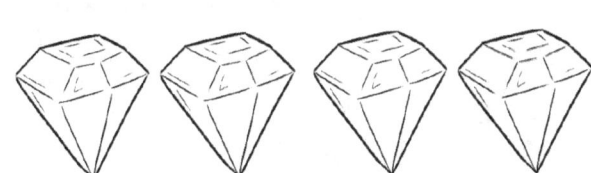 4-1=

 6-3=

SUBTRACTION

Count, Subtract and write the correct answer

$7-4=$ 3

$9-7=$ 2

$8-4=$

$9-5=$

SUBTRACTION

Count, Subtract and write the correct answer

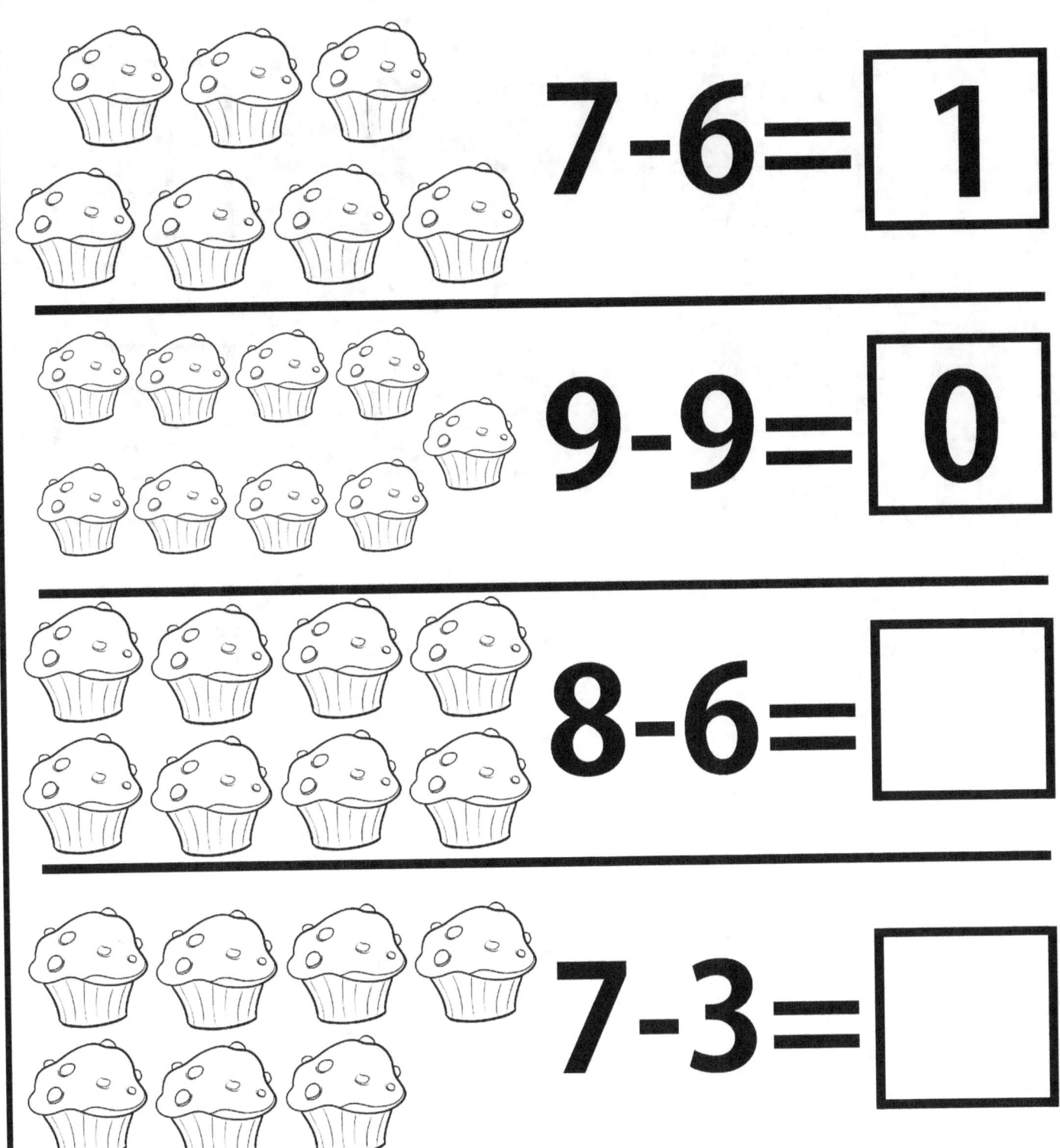

$7 - 6 = \boxed{1}$

$9 - 9 = \boxed{0}$

$8 - 6 = \boxed{}$

$7 - 3 = \boxed{}$

SUBTRACTION

Count, Subtract and write the correct answer

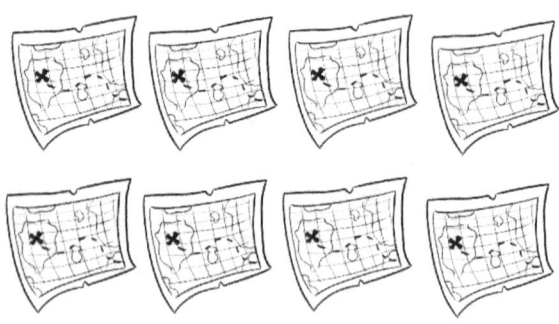 $8-2=\boxed{6}$

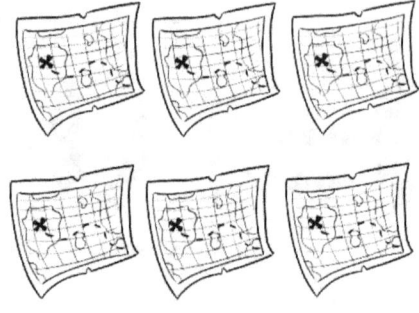 $6-2=\boxed{4}$

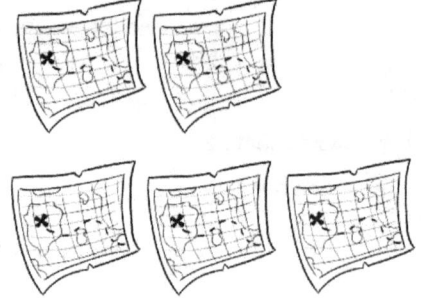 $5-5=\boxed{}$

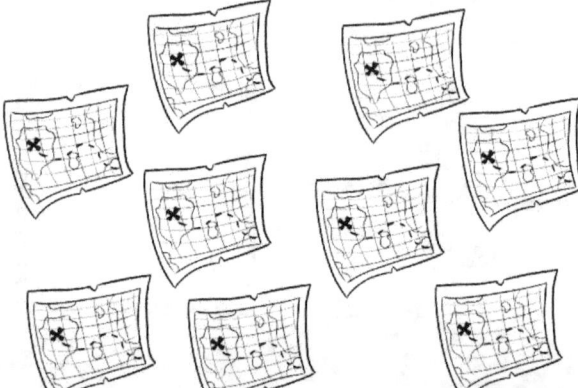 $9-3=\boxed{}$

BEFORE, BETWEEN AND AFTER

Fill in the box with the correct number

7 □ 9

2 3 □

□ 1 2

5 □ 7

BEFORE, BETWEEN AND AFTER

Fill in the box with the correct number

4      6

---

3   4   ☐

☐   2   3

---

6      8

BEFORE, BETWEEN AND AFTER

Fill in the box with the correct number

3 ☐ 5

8 9 ☐

☐ 6 7

4 ☐ 6

ADDITION

Choose the correct answer

ADDITION

Choose the correct answer

ADDITION

Choose the correct answer

🐼 **+** 🐼 **=** ☐

🐼🐼🐼🐼 **+** 🐼🐼🐼 **=** ☐

🐼🐼 **+** 🐼 **=** ☐

| 3 | 7 | 2 |

ADDITION

Choose the correct answer

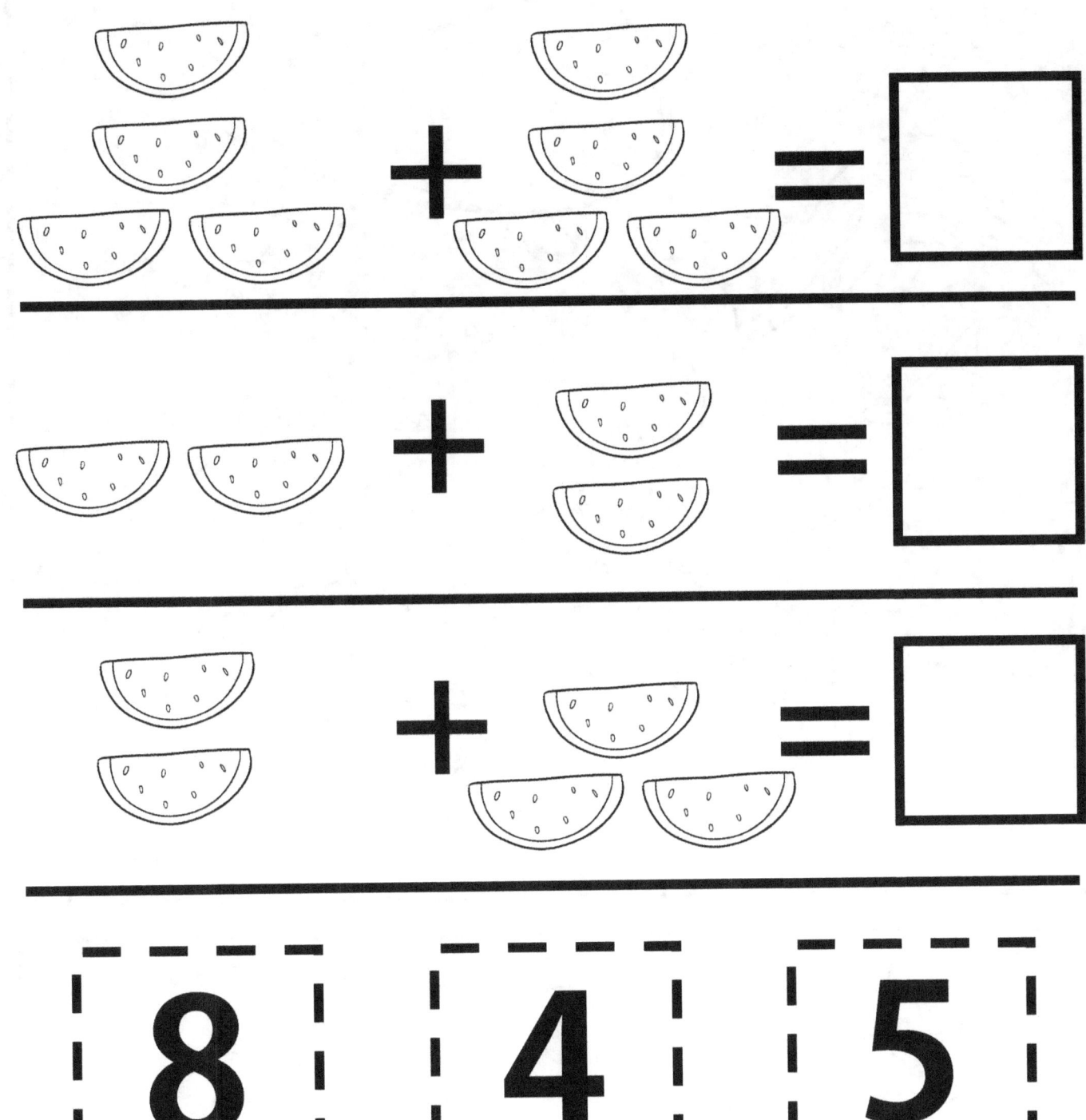

COUNTING

How many ?

 = **2**

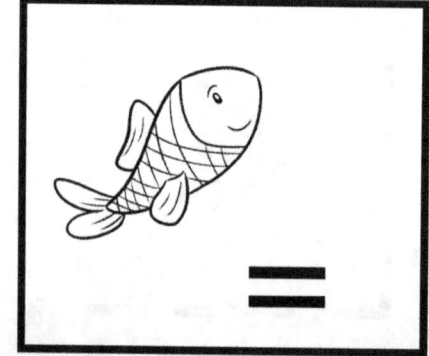 =

 =

How many ?

= 1

=

=

How many ?

 = 3

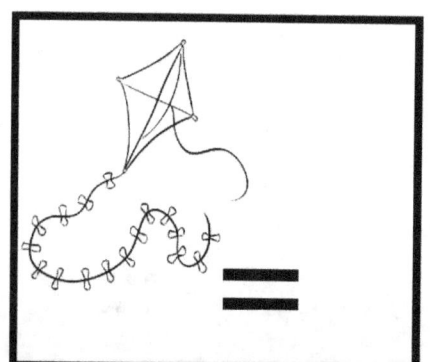 =

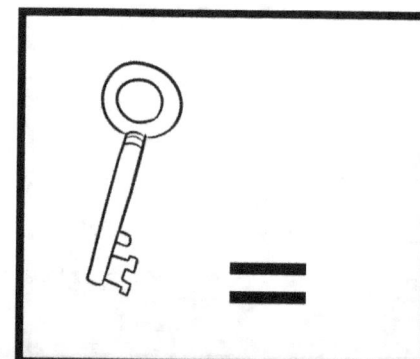 =

How many ?

 = 5

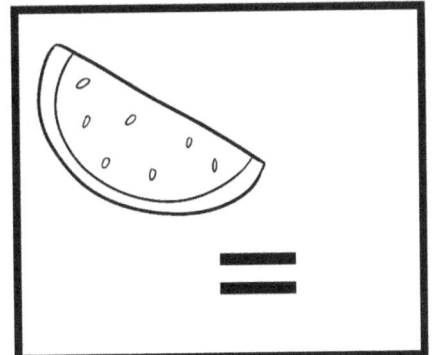 =

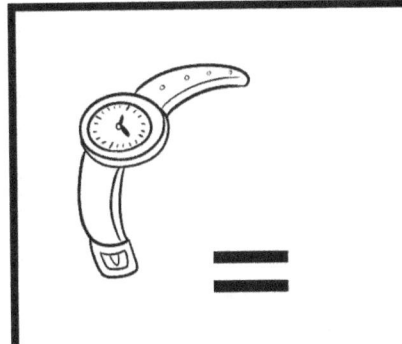 =

MISSING NUMBERS

Fill in the missing numbers

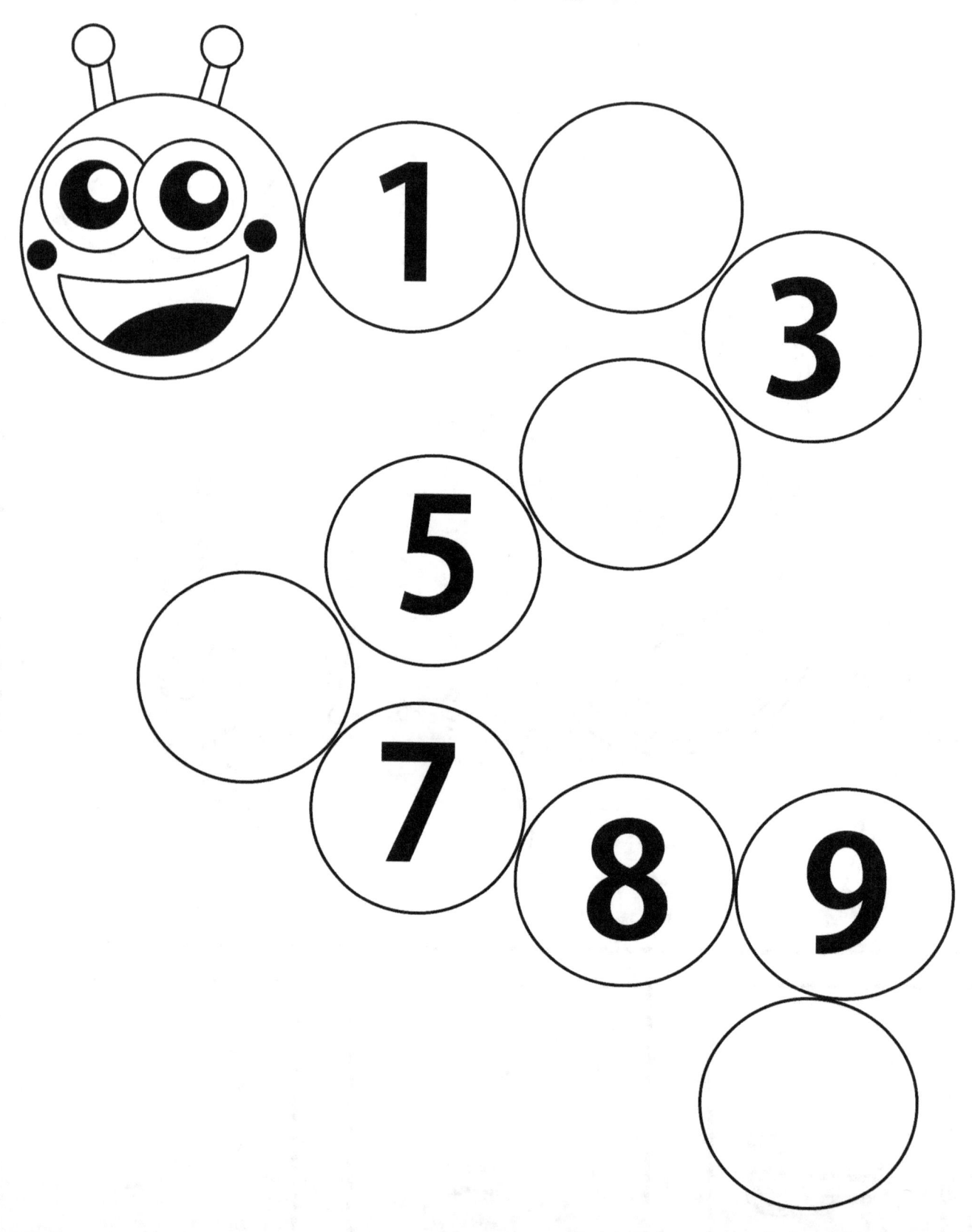

MISSING NUMBERS

Fill in the missing numbers